KNOWLEDGE ENCYCLOPEDIA
OUR UNIVERSE
SPACE

(An imprint of Prakash Books)

contact@wonderhousebooks.com

© Wonder House Books 2024

All rights reserved. No part of this book may be reproduced or transmitted in any form by any means, electronic or mechanical, including photocopying and recording, or by any information storage and retrieval system except as may be expressly permitted in writing by the publisher.

Disclaimer: The information contained in this encyclopedia has been collated with inputs from subject experts. All information contained herein is true to the best of the Publisher's knowledge. Maps are only indicative in nature.

ISBN : 9789390391264

Table of Contents

Our Amazing Universe	3
Studying the Universe	4
More about the Universe	5
Creation of the Universe	6–7
Matter & Antimatter	8–9
It's a Dark, Dark Universe	10–11
All about Galaxies	12–13
Two and a Half Galaxies	14–15
Three's Company: Quasars, Pulsars and Magnetars	16
Amazing Asteroids	17
Meteoroids, Meteors & Meteorites	18–19
Captivating Comets	20–21
Amazing Celestial Oddities: Black Holes & Auroras	22–23
Extrasolar Planets or Exoplanets	24–25
Universe or Multiverse?	26
Units of Measurement, and Cosmic Forces	27
Satellites: Natural & Artificial	28–29
Concepts about the Universe	30–31
Word Check	32

Our Amazing Universe

Have you ever looked up into the night sky and wondered what lies out there in the great universe? What really bewilders and fascinates you about the cosmos?

Today, we know that Earth is a very tiny ball of rock within the vast expanse of the universe. We also know that the birth of the solar system was perhaps only one amongst the many significant events that occurred in a fully developed universe. It is truly a humbling thought.

Wouldn't you like to know how the universe began or what galaxies, comets or shooting stars are? Read on to explore and understand some of the mysteries of our massive cosmos!

▼ *Stargazing is observing the stars in the night sky with or without a telescope*

The universe is a vast unending expanse of celestial matter and objects

Studying the Universe

All the matter and energy present everywhere, including in space, and life on Earth, is part of the mighty universe. Human beings have come a long way since ancient times when it was believed that the Sun, the Moon and Earth were the main celestial bodies of the universe. Today, we know that Earth and the solar system are only a tiny part of the vast universe.

What is the Universe?

Our universe is about 13.8 billion years old. Everything that exists in time and space—including all objects, energy, over hundred billion galaxies containing hundreds of billions of stars, the solar system, the planets, etc., are all part of what we call the universe. The universe is so vast that it really cannot be measured. We can only imagine how big it is by understanding that some celestial objects are so very far away in the universe that light travelling from them takes billions of years to reach Earth!

The universe is also referred to as the cosmos. In astronomy, cosmos is described as the entire physical universe, which is considered as one integrated whole. The word 'cosmos' comes from the Greek word, '*kosmos*' meaning 'order', 'harmony' and 'the world'. **Cosmology** is thus a particular field of study that combines natural sciences, specifically astronomy and physics, in a joint effort to understand the physical universe as a whole.

Space—on the other hand—is a limitless, three-dimensional expanse in which objects and events occur and have relative position and direction.

What Comprises the Universe?

Our universe is full of light-emitting and light-absorbing sources. It consists of stars, galaxies, quasars, clusters of galaxies and a complex cosmic structure. It also contains dust, neutral gas, dark matter, energy, different types of radiation and also black holes, amongst other cosmic bodies. It has vast empty spaces in between the structures.

Isn't It Amazing!

There are about a billion trillion stars in the universe. Some scientists believe that there are more stars in our universe than there are grains of sand on Earth!

▼ *Stars in the night sky*

More about the Universe

We know that the universe is vast and also that human beings do not really know its measure. In fact, the Milky Way galaxy, in which our solar system lies, is only one amongst the hundred billion galaxies which exist in the universe! The Milky Way is part of the Local Group—a group of over 20 or more neighbouring galaxies. With extensive research, we have been able to learn more about the composition and the shape of the universe.

Chemical Elements

Matter contains **chemical elements**. A chemical element is any substance that cannot be further broken down into simpler substances by ordinary chemical processes.

Hydrogen and helium are the main chemical components of the universe and were produced when the universe first came into existence. Some other 90-odd chemical elements are created in the stars, but these make up only a small percentage of the overall mass of the universe. These other elements are called 'metals' by astronomers, though in ordinary usage we do not refer to elements like oxygen and carbon as metals. The quantity of 'metals' varies depending on how stars were formed in that region.

Shape of the Universe

Scientists have concluded that mass causes space to curve. When objects move within that curved space, they are forced to alter or change their direction. If space is curved, then the shape of the universe may be any one of three types—flat, spherical or saddle-shaped. How significant are these shapes to understanding the universe?

The saddle-shaped surface has negative curvature. The flat surface is said to have zero curvature. The spherical surface has positive curvature.

If space has negative curvature, it means that the universe has no limits and will continue to expand forever. This is called an open universe.

If space has zero curvature, then the universe has no limits and will expand forever, but the expansion rate will gradually approach zero. This is termed a flat universe.

If space has positive curvature, it means the expansion will eventually stop and contraction will begin. So, galaxies will stop moving away from each other and get closer. As a result, the universe will disintegrate. This is a closed universe.

Recent observations show that the expansion of the universe is speeding up. This strongly implies that the universe is geometrically 'flat'. However, this still remains one of the major unexplained problems in modern cosmology.

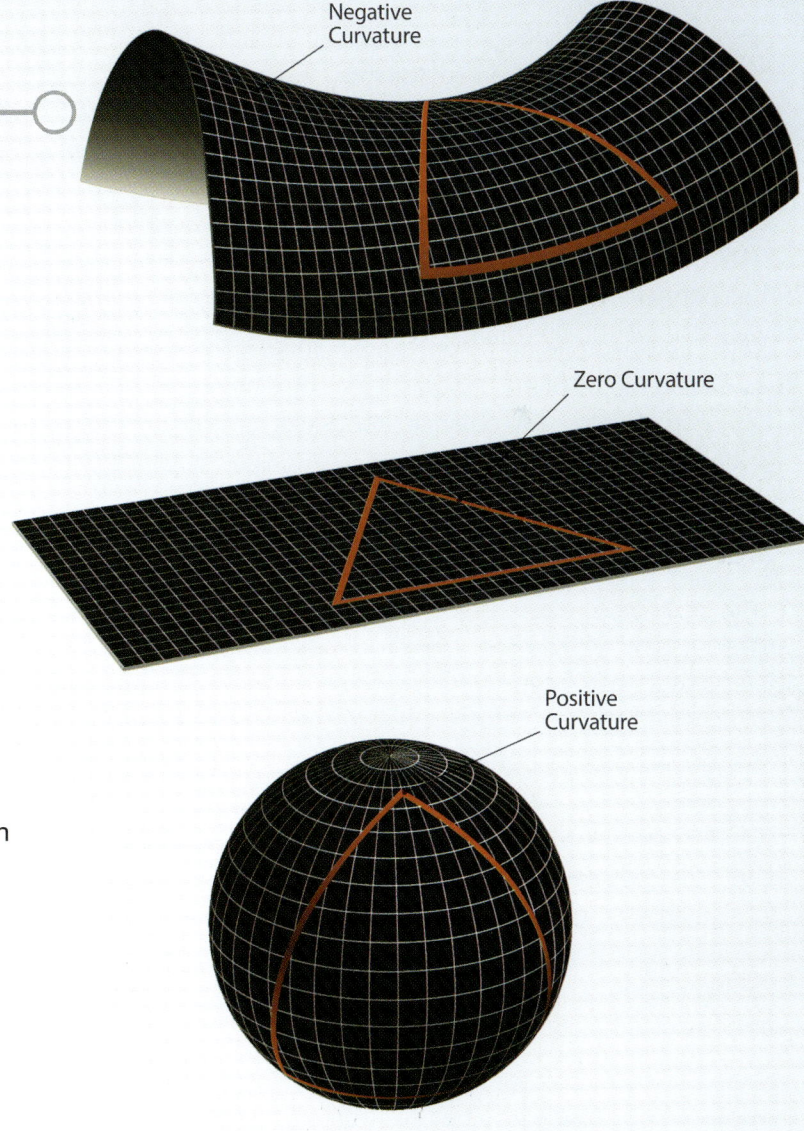

▲ Scientists believe that the destiny of the universe depends upon its shape or geometry. Some current observations of the speeding up of the expansion of the universe has led them to believe that the universe is flat

Creation of the Universe

What existed before the universe? How was it created and when? Nobody can really be certain of these answers, but the **Big Bang model** is considered to be responsible for this miraculous creation!

What is the Big Bang Model?

The Big Bang model is a theory about how the universe was formed and how it evolved. In fact, among the many theories on this subject, the Big Bang is the most popular. The essence of this theory is that the universe emerged from a state of very high temperature and density. This caused a violent collision on an extremely large scale, which is why the model is called the Big Bang. The theory was based on the observation that several other galaxies were moving away from the Milky Way galaxy in different directions and there seemed to be an ancient force responsible for this. Scientists estimate that the Big Bang happened nearly 13.8 billion years ago when the universe came into being.

Who thought of the Big Bang Model?

In 1927, a Belgian priest and astronomer, Georges Lemaître made a very important discovery—he independently proposed that the universe is expanding. In 1922, Russian mathematician Aleksandr Friedmann had also arrived at this conclusion. Lemaître claimed that the universe began as a single point, it had a finite beginning. It later expanded into its current vastness. He also said that it could keep growing. His formulation of the modern Big Bang theory was based on the work of Albert Einstein. Two years later, American astronomer Edwin Hubble confirmed Lemaître's theory that the universe was, in fact, expanding. He observed that galaxies were moving away from Earth and that the galaxies that were farther away were moving at a faster rate than those nearby. This meant that if things were moving away from each other, then perhaps a long time ago they existed close to each other.

While Dutch astronomer Willem de Sitter had also earlier considered an expanding model of the universe, Lemaître's theory, which was revised by George Gamow and others in the 1940s, remains the principal model of the universe.

▲ *Aleksandr Friedmann*   ▲ *Georges Lemaître*

The Theory

According to the Big Bang Theory, the universe started off as a hot and extremely dense point, about a few millimetres wide. Approximately 13.8 billion years ago, this tiny point (or **singularity**) exploded in a violent bang from which all of matter, energy, space and time were created.

Different theories state that immediately after the Big Bang, there probably was a colossal sea of protons, electrons, neutrons and other particles. With time, the universe continued to cool, resulting in the decaying of some particles and the recombining of some. Protons and electrons may have combined to form neutral hydrogen. The universe may have been opaque before the recombining, due to the scattering of light by the free electrons. Once neutral **atoms** were formed, it became transparent. The atoms joined together and after a very long time formed stars and galaxies. The initial few stars were responsible for creating bigger atoms and also for groups of atoms known as **molecules**. This led to the birth of many more stars. Simultaneously, galaxies were banging against each other and coming together. During this process of formation and dying of stars, things like asteroids, comets, planets and black holes were created.

Since the theory predicted that the early cosmos was in a very hot state and that the gases cooled with expansion, it would most probably also be filled with radiation or remains of heat from the violent explosion. The remains of this hot dense matter are called **cosmic microwave background** (CMB) radiation. CMB radiation today is very cold and is similar to what is used to emit TV signals through antennae. A lot of information about the early universe can be gathered from CMB, the earliest known form of radiation. CMB was first predicted by Ralph Asher Alpher in 1948, when he was doing research about the Big Bang model along with Robert Herman and George Gamow.

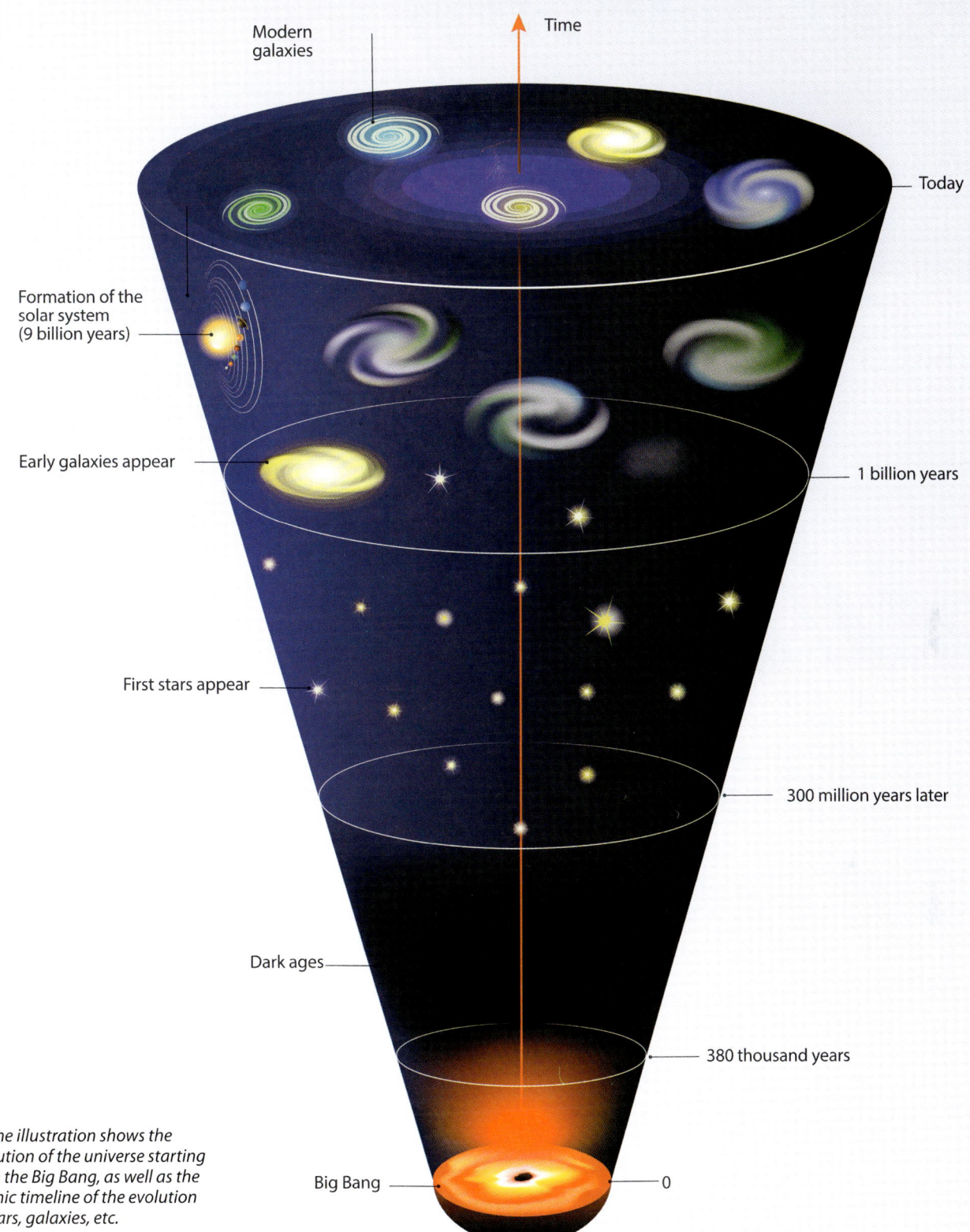

▶ The illustration shows the evolution of the universe starting from the Big Bang, as well as the cosmic timeline of the evolution of stars, galaxies, etc.

⭐ Incredible Individuals

While Friedmann and Hubble proposed the Big Bang theory, it was George Gamow who took up the task of coming up with a practicable model from this theory. Though Friedmann's theory was brilliant, it was not known by many. Those who had heard it did not believe it. But Gamow and Alpher had faith in the theory. At the time, Ralph Alpher was a mathematics student who was working on describing how the universe evolved. CMB was first noticed accidentally, in 1965, by Arno Penzias and Robert Wilson at the Bell Telephone Laboratories in New Jersey, USA. So, the Big Bang model was the brainchild of several people.

Matter & Antimatter

In our universe, anything that has mass, takes up space and has volume is called matter. This means that ordinary or regular matter is something we can see, feel, touch and even taste. Antimatter is the opposite. It is matter made up of antiparticles that have properties opposite those of normal matter. Antiparticles are subatomic particles having the same mass as a given particle, but with an opposite electric charge. To understand antimatter, let us first take another look at what matter is.

What is Matter?

Everything around us—all solids, liquids and gases comprise atoms which are the main ingredients of all matter. At the centre of each atom is a nucleus which comprises two different particles—protons and neutrons. The nucleus is surrounded by smaller particles called electrons. So basically, all matter is made up mainly of neutrons, protons and electrons. Both protons and electrons have an electric charge. Protons are positively charged, and electrons are negatively charged.

When there are two opposite charges—negative and positive—they attract each other, but when the charge is the same, they repel each other. Suppose you rub a plastic ball against your hair, and you see your hair rising up, this is due to the charge. When you rub them against each other, the protons and electrons get unevenly distributed over the ball and your hair. This results in the ball getting either positively or negatively charged and it then sticks to anything which has an opposite charge.

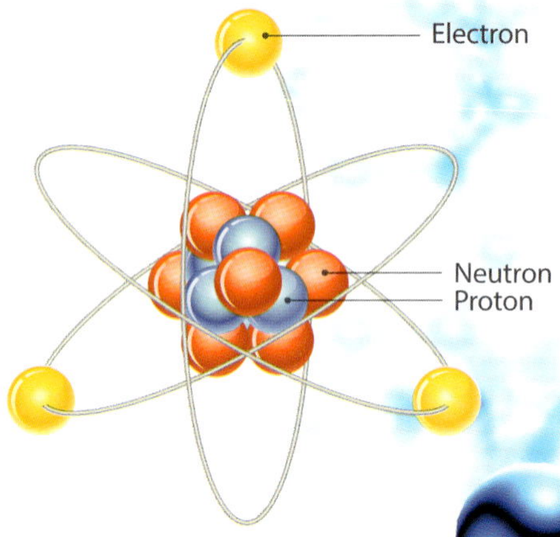

▲ An illustration of the structure of an atom made up of protons, neutrons and electrons. Protons and neutrons make up the nucleus of the atom

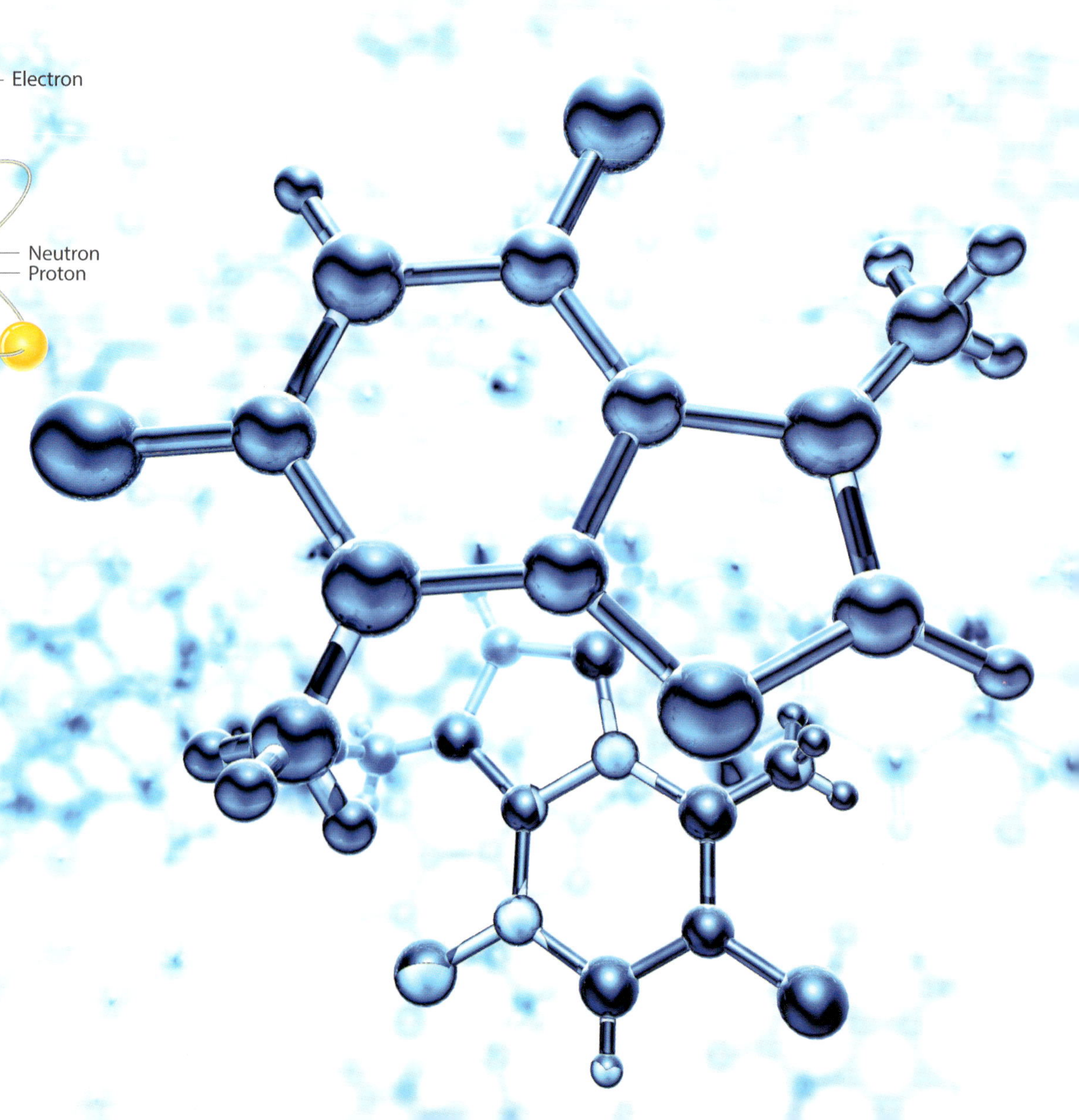

▶ The illustration shows the chemical structure of molecules. Molecules are small structures (of two or more atoms) that make up matter

▲ An artist's rendering of flying antimatter particles reacting with matter particles in a nebula

What is Antimatter?

Antimatter is almost the same as regular matter, but it comprises antiprotons, antineutrons and antielectrons or positrons. The main difference between the particles of matter versus antimatter is that, in the latter, the corresponding antiparticles have the reverse charge. Anti protons, therefore, are negatively charged, and antielectrons have a positive charge.

So, what happens when particles and antiparticles collide with each other? When an electron and a positron come into contact with each other, the two get destroyed, leaving behind gamma rays or radiation. To put it another way, the mass of particles is converted into pure energy. The energy released during this process or collision will be equal to the mass of the two particles into twice the speed of light. It is a huge amount of energy. Matter and antimatter, therefore, cannot exist close together for more than a small fraction of a second because they will crash against each other and release energy.

Antimatter exists in the universe, but there is not much of it. Cosmic rays found in outer space are a natural source of antimatter. Another source is radioactive decay. Antimatter can also be created artificially in a science laboratory.

Scientists have not yet been able to figure out why there is so little antimatter in our universe in comparison to normal matter and also why they both do not exist in equal amounts.

Isn't It Amazing!

Hydrogen atoms are available in plenty in the universe. These atoms are also the simplest of all the other elements in the universe. The very first anti-atom, or antimatter counterpart of a regular atom, was created in 1995 by physicists at CERN—the European Organisation for Nuclear Research, in Geneva. It was the anti-hydrogen atom. Unlike hydrogen, anti-hydrogen is rare and difficult to produce.

▲ The Globe of the Science and Innovation Centre at CERN in Geneva

It's a Dark, Dark Universe

Human beings have made great scientific progress in all fields including the exploration of space and understanding the universe. Yet, you may be surprised to know that most of the universe remains a mystery to us. Why is that so? More than 95 per cent of the universe is made up of energy and matter that nobody has been able to understand or study till date.

★ Dark Energy

Stars, planets, black holes, asteroids, comets, etc., are only a small fraction of what makes up the universe. Based on their observations and studies of these objects, scientists have concluded that our universe is in a state of expansion. But if it was only made up of galaxies, stars, planets and other objects that we know about, then it should not be expanding. This contradiction led to the belief that something more—or rather some other energy—exists in the universe, which is causing it to expand. This is called dark energy.

While not much is known about dark energy, we are sure that there is a lot of it present in the universe. In fact, it makes up 68 per cent of the universe.

◀ A collage compiled from images taken by the NASA/ESA Hubble Space Telescope shows six different galaxy clusters. The images map the post-collision distribution of stars and also of dark matter (coloured in blue)

▼ The California nebula in the constellation of Perseus, with the bright star Menkib

Dark Matter

Besides dark energy, there is another source of gravity in space. Scientists have been able to observe its pull upon objects like stars, galaxies and other celestial objects. Yet, it is not what we normally know or recognise as matter, neither is it a black hole. It is something quite different. While we know that it exists, nobody knows what it really is. Scientists have termed this stuff as dark matter. It is a part of the universe whose presence is felt due to its gravitational attraction and not its intrinsic brightness. 27 per cent of the universe is made up of dark matter.

So, this means that over 90 per cent of the universe consists of dark energy and dark matter, both of which are largely unknown to us. All the rest of the visible and ordinary (or **baryonic**) matter, like energy, light, heat, X-rays, human beings, the solar system, galaxies, etc., makes up 5 per cent of the universe! That is a very small part of the universe which we have been able to observe, study and figure out.

Dark energy and dark matter are important in the study of space and physics. Scientists are trying to understand them through observations and mathematical calculations. Eventually, this will provide us with information about our amazing universe.

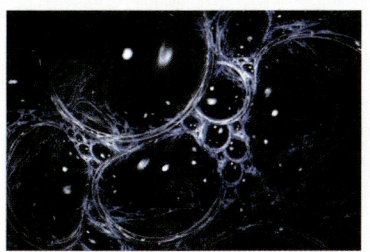
▲ An artist's rendering of space closed in energy tension, which is the basis for dark matter and dark energy

▲ This composite image shows a ring of dark matter seen in a galaxy cluster. The image was taken by NASA's Hubble Space Telescope. Credit: NASA, ESA, M.J. Jee and H. Ford (Johns Hopkins University)

Who Discovered Dark Matter?

Dark matter, originally known as 'missing mass', was first worked out by Swiss-American astronomer Fritz Zwicky in 1933. While studying the mass of stars in the Coma cluster of galaxies, Fritz found that the stars consisted of only about 1 per cent of the mass required to keep the galaxies together. But there was other mass which existed. This was responsible for keeping the clusters together and providing the extra gravitational pull required to hold galaxies together. For several years, the mystery of this missing mass was not solved. In the 1970s, Vera Rubin and William Kent Ford (both American astronomers) confirmed the existence of dark matter by observing a similar occurrence.

◄ Fritz Zwicky ► Vera Rubin

All about Galaxies

Galaxies are home to stars and other celestial objects. It was only in the early 20th century that the existence of galaxies other than the Milky Way was recognised. Before that, early astronomers labelled them as nebulas, since they appeared to look like hazy clouds. Galaxies are present in every part of space, as observed through powerful telescopes. They differ in shape, structure and the level of activity within them.

What is a Galaxy?

A galaxy comprises a large group of hundreds of billions of stars and **interstellar** matter (gas and dust) bound together by gravity. Almost all the large galaxies are also believed to have gigantic black holes at their centres. Galaxies exist in a variety of shapes and sizes ranging from dim dwarf-sized objects to bright, massive spiral-shaped ones. Almost all galaxies seem to have been formed immediately after the universe came into existence. These beautiful formations are generally found in clusters, some of which form a larger cluster and span hundreds of millions of **light years** across the universe. A light year is the distance travelled by light in one year, at a speed of 3,00,000 kmps.

Types of Galaxies

There are three main classifications of galaxies—elliptical, spiral and irregular, as seen in the diagram. Some spiral galaxies are called 'barred spirals'.

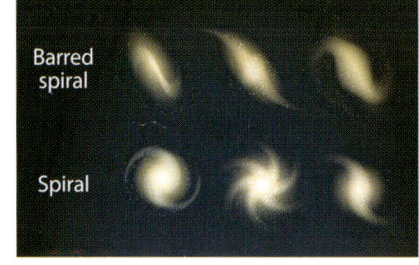
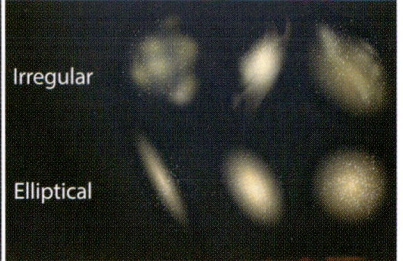

▶ Types of galaxies

Elliptical Galaxies

These galaxies are round, oval or more like an elongated sphere. Sometimes, they may be so stretched that they look like a cigar. These galaxies generally contain many old stars, but not much dust and other interstellar matter. Like the stars in the discs of spiral galaxies, their stars also orbit around the galactic centre, but in a more disorderly way. Not many stars have been known to have formed in elliptical galaxies. The largest known galaxies in the universe are giant elliptical ones which can be as big as two million light years long. The smaller of these galaxies are known as dwarf elliptical galaxies. Virgo A (or M87) is an example of a giant elliptical galaxy found close to the centre of the Virgo cluster of galaxies.

▲ Seen here are the dust lanes and star clusters of the NGC 1316 giant elliptical galaxy in the Fornax constellation

▲ The Messier 59 is an elliptical galaxy. It is also one of the largest elliptical galaxies in the Virgo galaxy cluster

Spiral Galaxies

Spiral galaxies comprise a flat disc with a bulging centre. These galaxies have long spiral arms that wind towards the centre. The disc comprises stars, planets, dust and gas, which rotate or spin around the galactic centre in a regular manner, at hundreds of kilometres per second. This spinning motion may result in matter in the disc taking the shape of a spiral, like a pinwheel.

While many new stars are born in spiral galaxies, the older stars are generally located in the bulging centre of the galactic disc. These discs have a halo around them and astronomers believe that they comprise unknown dark matter. The Milky Way is a spiral galaxy and is home to our solar system. It is one from a group of galaxies known as the Local Group.

One type of spiral galaxy is known as 'barred spirals' since the bulge at the centre looks stretched like a bar and the spiral arms come out from the ends of the bar.

◀ An infrared picture of M74—a spiral galaxy—as seen by NASA's Spitzer Infrared Array Camera

▶ A barred spiral galaxy—NGC 1672 in the constellation of Dorado

 # Irregular Galaxies

Galaxies which do not have any distinct shape, such as spiral, elliptical or lenticular (resembling lenses), are irregular galaxies. Irregular galaxies—like the Large and Small Magellanic Clouds—are uneven or out of shape as they are generally under the gravitational influence of other nearby galaxies. Since they are packed with lots of gas and dust, irregular galaxies are a fertile ground for the formation of new stars.

◀ An image of an irregular galaxy—NGC 1427A

Credit: NASA, ESA, and The Hubble Heritage Team (STScI/AURA)

In Real Life

One of the most important things needed to support and sustain life on Earth and in the universe is water! Water in its solid state, as ice, exists in abundance in the universe and is found in interstellar dust clouds as well as in the orangish-red fields of Mars. However, that by itself is not enough to support life. Water, in its liquid state, acts as a crucial lubricant for the molecular or chemical processes for all forms of life like human beings, plants and animals. Hence, astronomers always look for signs of water in its liquid state in the universe to see if alien life exists elsewhere.

 ## Incredible Individuals

In 1950, Arthur Allen Hoag, an American astronomer, discovered one of the rarest kinds of galaxies, a type of ring galaxy consisting of a symmetrical central core made up of older stars, surrounded by a bright ring of young blue stars with no apparent connection between the two. It came to be known as the Hoag's Object. These rare galaxies comprise less than 0.1 per cent of all observed galaxies in the universe.

Two and a Half Galaxies

The Milky Way galaxy is one amongst the trillions of galaxies found in our universe. However, it is the most important galaxy for us because our solar system lies within it. The Andromeda is another important galaxy, since it is the closest large galaxy to the Milky Way, and it is one of the few that can be seen with the naked eye. The Milky Way also has two satellites or companion galaxies known as the Magellanic Clouds.

⭐ The Milky Way Galaxy

The Milky Way galaxy, named so due to its 'milky' appearance, contains several hundred billion stars. It roughly spans about a hundred thousand light years across and is about a thousand light years in thickness. The solar system (comprising our Sun and its planets) is located in the curved arm of gas and dust of the galaxy, and it is approximately 26,000 light years away from its raging centre.

It has a supermassive black hole at its centre and is surrounded by dark matter bigger than the galaxy itself. Even though Earth lies within the Milky Way, astronomers do not know as much about this supermassive black hole as they do about some of the other celestial objects in our galaxy. This is due to a thick layer of dust which prevents optical telescopes from getting a good view.

▲ *An infrared image taken by NASA's Spitzer Space Telescope reveals the millions of stars clustered together in the centre of the Milky Way galaxy. An ordinary optical telescope would not have been able to view this*

◄ *Our home galaxy, the Milky Way*

⭐ Andromeda: The Gobbling Galaxy

The Andromeda galaxy (or M31) is a pancake-shaped, disc-like spiral galaxy. It is a large galaxy nearest to the Milky Way. It was first mentioned in 965 CE in *The Book of the Fixed Stars*.

A long time ago, there existed three large galaxies—Andromeda, the Milky Way, and its smaller sister galaxy called M32p. These three galaxies circulated close around one another and gobbled up matter and other smaller galaxies. Much later, Andromeda collided into M32p, shredding and devouring it, and leaving behind an almost invisible halo of stars.

Andromeda has some unusual features: a dim halo or ring-like pattern of stars orbiting it and a small and dense satellite galaxy called M32.

After the telescope was invented in 1611, German astronomer Simon Marius discovered the Andromeda galaxy. For a long time, Andromeda was considered to be a part of the Milky Way. In 1924, Edwin Hubble concluded that Andromeda was actually a separate galaxy.

▲ *The Andromeda galaxy against the Milky Way (elements of this image are furnished by NASA)*

Incredible Individuals

Turkish-born astrophysicist Burçin Mutlu-Pakdil, who is now a postdoctoral research associate at the University of Arizona's Steward Observatory, USA, has a remarkable achievement she can be proud of. She discovered a rare and unknown type of galaxy more than 350 million light years away, which now bears her name and is known as Burçin's galaxy! Being a female astrophysicist in a male-dominated field was not an easy task. While her family encouraged her to fulfil her dreams, she came from a society that shunned women for moving cities in order to study. In spite of numerous such hurdles, she persevered with her passion.

◀ Dr Burçin Mutlu-Pakdil developed a love for astrophysics while she was still in middle school. She was one of only 20 change-makers from all over the world who were selected and invited to become a TED fellow in 2018

▲ Turkish astronomer, Dr Burçin Mutlu-Pakdil at the Subaru Telescope

▲ This photograph was taken at the site of the European Southern Observatory's Very Large Telescope in the Chilean Atacama Desert. It shows the Large and Small Magellanic Clouds glowing brightly on the extreme left. Photo by ESO Photo Ambassador Yuri Beletsky

Satellite Galaxies—The Magellanic Clouds

Besides the Sun and the other numerous stars, there are smaller galaxies comprising their own group of stars, which also orbit around the Milky Way and are termed as satellite galaxies.

The biggest are the Magellanic Clouds, also known as Nubeculae Magellani—two unevenly shaped galaxies which can be seen in the Southern Celestial Hemisphere—named after Portuguese navigator Ferdinand Magellan. He and his crew discovered them during the first voyage around the world (1519–1522). These companion galaxies, however, were properly recognised only in the 20th century by Edwin Hubble.

▶ The Large Magellanic Cloud is also referred to as a satellite or dwarf galaxy. The image was captured in infrared light as observed by the Herschel Space Observatory and NASA's Spitzer Space Telescope

Three's Company
Quasars, Pulsars and Magnetars

There are some celestial objects that are lesser-known. That is because they are either too far away from us and difficult to study, they can only be observed under certain conditions or because of their strong magnetic fields.

Quaint Quasars

A quasar (pronounced *kwayzar*) is an extremely bright astronomical object similar to a star. It can be a trillion times brighter than our Sun! Quasars give off large quantities of energy. This energy is derived from gigantic black holes which exist in the centre of galaxies where quasars are located. Quasars are very bright. While they outshine all other stars in the same galaxy, they cannot be seen by the naked eye as they are amongst the farthest objects in space. Most quasars are larger than our solar system and emit more energy than 100 normal galaxies combined!

▶ A bright quasar

 Pulsating Pulsars

Neutron stars that spin extremely fast and have a regular pulse of **radiation** or **radio waves,** are known as **pulsars** or pulsating radio stars. A neutron star is born when the centre of a star explodes violently in what is known as a supernova. Since a neutron star has a solid core and a liquid mantle, it has a magnetic field a trillion times stronger than Earth's. It emits high-energy beams at both the North magnetic pole as well as the South. As the neutron star rotates, if at any time a beam point towards Earth, it appears to turn on and off, to pulse! We can therefore conclude that all pulsars are neutron stars but all neutron stars are not pulsars.

A pulsar is like a lighthouse. Although its light is shining all the time, we can only see the beam of light when it points directly in our direction. Similarly, a beam of light from a pulsar can be seen intermittently only when it crosses our line of vision.

▲ An illustration of a pulsar, a highly magnetised rotating neutron star

Magical Magnetars

A magnetar is another kind of a neutron star. We know that a regular neutron star has a magnetic field which is a trillion times that of Earth's magnetic field. The magnetic field of a magnetar is additionally 1000 times stronger. Magnetars are the most powerful magnets in the universe. It is said that if a magnetar did actually come as close as about 966 kilometres to Earth, the pull would be so strong that it could possibly suck out iron from our bodies!

 Isn't It Amazing!

In December 2004, NASA, European satellites and other radio telescopes observed a powerful flash of light from across the Milky Way galaxy. The light was so strong that it deflected off the Moon, causing Earth's upper atmosphere to light up. The light came from a magnetar called SGR 1806-20 which blew up, and in a tenth of a second it released huge amounts of energy, more energy than the Sun has released in 1,00,000 years!

▶ The SGR 1806-20 magnetar with magnetic field lines, as conceived by an artist

Amazing Asteroids

Besides the planets, the Moon and the stars, there are several smaller celestial objects in space called 'small bodies'. These include asteroids, meteoroids and comets.

What is an Asteroid?

An asteroid is a small, rock-like object which orbits the Sun, just like planets. But unlike planets, asteroids are much smaller. They are sometimes called minor planets.

A majority of the several hundred thousand asteroids in our solar system are found in the asteroid belt. It is a flat ring-like region between Mars and Jupiter. Some asteroids are found in the orbital path of planets like Earth.

Where did Asteroids Originate From?

Asteroids are leftover pieces from the time when the solar system was formed around 4.6 billion years ago! The solar system began with the collapse of a huge cloud of gas and dust, which condensed to form the Sun, planets and their moons. Asteroids are the remains in the asteroid belt that never made it to the Sun, nor could they transform into planets or any other celestial bodies.

Are all Asteroids Identical?

Asteroids are different from each other and were created in different places and at different distances from the Sun. Most have a sharp and uneven shape. Some asteroids are hundreds of kilometres in diameter, but many are the size of a small stone. Asteroids are made up of various types of rock, clay or metal, like nickel and iron. They provide important information regarding our planets and the Sun.

In Real Life

Many space missions have been sent by NASA to observe asteroids.

2001: The NEAR Shoemaker spacecraft went to Eros, an asteroid near Earth.

▲ The Eros asteroid

2011: The second-largest object in the asteroid belt—Vesta, a small planet, was orbited and studied by the Dawn spacecraft.

◄ Vesta

2012: Dawn orbited and studied the dwarf planet Ceres, the largest object in the asteroid belt.

► Ceres was the first asteroid to be discovered in 1801

2016: The OSIRIS-REx spacecraft was launched by NASA to study Bennu, which is an asteroid near Earth. The objective was also to bring a sample of it back to our planet for study.

◄ An artist's conceptualisation of NASA's OSIRIS-REx spacecraft

Meteoroids, Meteors & Meteorites

Meteoroids, meteors, and meteorites are all related to the 'shooting stars' we sometimes see streaking across the night sky. We call the same celestial objects by different names, depending on where they are.

⭐ What are Meteoroids and Meteors?

When one asteroid bangs into another, it may break into pieces—these pieces are called meteoroids. A meteoroid is a small rocky or metallic natural object that enters Earth's atmosphere. When a meteoroid falls to Earth with great speed, there is a resistance (or drag) of the air on the rock, which heats it up.

As it falls and comes closer to Earth and passes through our atmosphere, it starts to vaporise (becomes gaseous), and a streak of light is seen, which is the hot air left behind by the burning piece of rock. This is a meteor, a streak of light in the sky. Meteors are not really stars, but due to their appearance and streaks of light, they are also known as 'shooting stars'! Meteors are sometimes confused with comets due to the light they both seem to emit. However, they are different from comets. Comets are made of ice and dust, not rock.

⭐ Meteorites

Most meteoroids get vaporised by the time they enter Earth's atmosphere, however some of these rocks do not disintegrate. Instead, they reach the surface of Earth and are known as meteorites. Most meteorites are the size of a small pebble, but some rare ones can also be the size of a large boulder. Since meteorites originate from asteroids, they are useful to scientists, who can gain more information about these ancient rock-like objects.

Asteroid — Space

Meteoroid

Meteor — Earth's atmosphere

◀ *The illustration shows the difference between asteroids, meteoroids, meteors and meteorites*

Meteorite — Earth's surface

👤 In Real Life

The Hoba is the largest meteorite found on Earth. Found in 1920 in Namibia, Africa, it weighs approximately 53,977 kilograms! It is an ataxite, an iron meteorite which contains more than 16 per cent nickel.

▼ *The largest meteorite, Hoba was found in Namibia*

SPACE | OUR UNIVERSITY

▲ Meteor showers are fascinating to observe

⭐ It's Raining Meteors!

When many meteors fall to Earth at the same time, they are referred to as a meteor shower. Meteors fall at a speed which is 32 times faster than that of a speeding bullet!

A meteor shower is generally named after the constellation from which it appears to be coming. Scientists have estimated that there are nearly 21 meteor showers that occur annually. Listed below are some of the major meteor showers, their constellations, and the months when they can be viewed.

Quadrantids (originally Quadrans Muralis, now Bootes constellation): December/January

Lyrids (Lyra constellation): April

Perseids (Perseus constellation): August

Orionids (Orion constellation): October

Leonids (Lyra constellation): November

Geminids (Gemini constellation): December

💡 Isn't It Amazing!

More than 45,000 kilograms of space debris falls on Earth every day. Meteors enter Earth's atmosphere at high speeds ranging from over 40,000 kilometres per hour to 2,57,495 kilometres per hour—unimaginable to the human mind!

In the year 1908, an object as large as a residential building fell from the sky and exploded in the air above Siberia. Known as the Tunguska event, named after a river, this object razed trees in an area spanning nearly 2,072 square kilometres. Luckily, no human being or creature was killed or hurt, but it is one of the most significant events of this kind to ever be recorded in human history. Scientists are not sure of the object's origins and whether it was a comet or an asteroid.

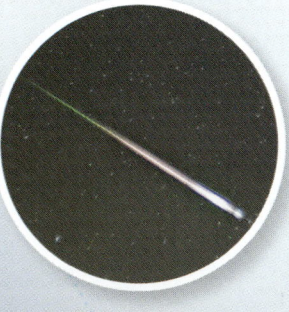

▲ A Siberian meteorite

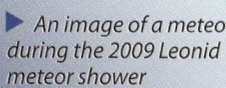

▶ An image of a meteor during the 2009 Leonid meteor shower

Captivating Comets

People sometimes confuse a comet with a meteor. However, they are both different in their composition. Besides, unlike a meteor, comets can be viewed even when they are very far from Earth.

What is a Comet?

Comets, also called 'dirty snowballs', are a part of the solar system, and are typically icy bodies covered with dark organic material. Like asteroids and planets, they also orbit the Sun.

However, comets have a very long orbit. When a comet orbits too close to the Sun, the ice and dust begin to get destroyed, or vaporise. This creates a cloud of comet dust particles around the heart or nucleus of the comet and is known as the coma (see diagram below). The vaporised ice and dust form the tail of the comet. The tail glows due to light from the Sun, and is visible to human beings on Earth.

▼ A comet streaking across the sky

Comets generally have two different types of tails—the white ones are made up of comet dust and the bluish ones of electrically charged gas.

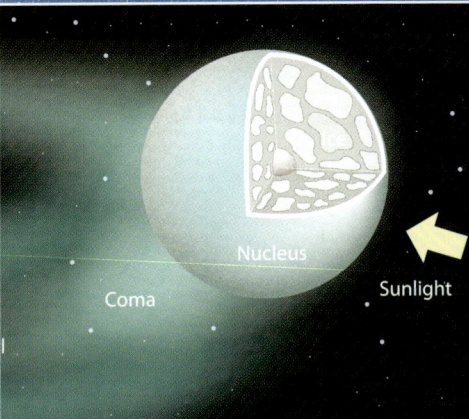

▲ The diagram shows the structure of a comet with its nucleus, coma, dust and gas tails

Incredible Individuals

Caroline Herschel was the sister of a well-known German astronomer named William Herschel. Caroline came from a family of musicians. She was a singer, a mathematician and an astronomer. While her father encouraged all his children to study French, mathematics and music, her mother did not consider it important to educate a girl. She made her do all the housework. At age 10, Caroline became sick and disease stunted her growth. She grew no taller than 4 feet 3 inches. In her early adulthood, she lived with her brother William in England, who trained her in singing, mathematics and astronomy. She often helped him with his work. She learned how to grind glass to make powerful telescopes and began using them to make her own astronomical discoveries.

Three new nebulas (the region where stars are born) were discovered by her. Caroline was the first woman to find a comet; in fact, before she died, aged 98, she had discovered eight of them!

▶ Caroline Herschel (1750–1848) was the first woman to find a comet

SPACE — OUR UNIVERSE

▲ The two main regions or belts from which comets originate—the Kuiper Belt and the Oort Cloud

⭐ Chasing Comets

While billions of comets can be found in the solar system, most of these originate from two regions—the Kuiper Belt (a ring of icy bodies lying outside Neptune's orbit) and the Oort Cloud (a ring-like region beyond Pluto, on the outer edges of the Kuiper Belt).

Bright comets which are visible on Earth at night appear approximately once every 10 years. Short-period comets, like the Halley's Comet, traverse through the solar system maybe once or twice during a human lifetime. The Halley's Comet takes less than 200 years to orbit the Sun. In 2061, it will return on its regular 76-year journey around the Sun. Long-period comets from the Oort Cloud pass close to the Sun only once in every 100 to 1000 years, and they are less predictable.

Comets generally travel at a safe distance from the Sun, like the Halley's Comet. However, some crash headlong into the Sun or get so close to it that they break up and evaporate—these are known as sungrazers. Comets are generally given names depending on who discovered them, either a person or a spacecraft.

▲ A picture of the Halley's Comet taken on June 6, 1910

👤 In Real Life

Astronomers now have a much better idea of what makes up a comet's nucleus. NASA's Deep Impact spacecraft's 'smart impactor' blasted out a huge crater from the nucleus of the comet Tempel 1 in July 2005. They found the nucleus to be spongy with lots of holes; some sections of the surface were delicate and weak; the surface of the nucleus was covered with fine dust.

▶ Halley's Comet is named after English astronomer Edmond Halley. It is predicted to return to Earth every 76 years. It was last seen in 1986. It will now enter the inner region of the solar system only in 2061. Keep your eyes peeled!

Amazing Celestial Oddities
Black Holes & Auroras

The mysteries of the universe are unending. Here are two more that are mindboggling; one is the blackest object to be found in deep space, and the other is a celebratory vibrant show of lights seen on Earth!

 ## What are Black Holes?

What can be darker than space itself? Well the darkest objects to be found in the depths of our universe are also some of the weirdest and strangest things—black holes! A black hole is an area with such tremendous gravity or pull that nothing—not even light—can escape from it. Usually, black holes are formed when a star dies. When the fuel in a star gets depleted or finished, it starts to disintegrate and cave in on itself, resulting in a huge bang. All the matter remaining after the explosion—which is much, much more than the mass of the Sun—falls into a really tiny point. This point where a large amount of mass is trapped is called a singularity. It has a huge impact even though it is small.

You can imagine a black hole as a circle with a singularity in the centre. The gravity in this centre is so great that it sucks in everything, even light. That is why black holes appear so black! The circle is known as the event horizon or a black hole.

▶ The first-ever image (see inset) of a supermassive black hole and its shadow in the centre of Messier 87 (a gigantic galaxy in the Virgo cluster) made headlines in April 2019. It was captured by the Event Horizon Telescope (EHT)—an international collaboration of radio telescopes

 ## The Aura of Auroras

If you are lucky enough to find yourself near the North Pole or the South Pole, you may want to check out the amazing and exquisite light shows in the sky. These lights are known as auroras. The lights near the North Pole are called Aurora Borealis or the Northern Lights and the ones near the South Pole are referred to as Aurora Australis or the Southern Lights.

▲ Brilliant green Northern Lights at Lake Laberge, Yukon, Canada

◀ The Northern Lights during winter in Tromso, Norway, in front of a *fjord*

What Causes Auroras?

Auroras are mostly seen during the night, but they are effects created by the Sun. Besides heat and light, the Sun is responsible for sending us small particles and a lot of other energy. Earth's **magnetic field** protects us from this energy and particles.

Auroras are caused by charged particles that travel between the Sun and Earth along magnetic fields. A magnetic field is the area of influence of a magnet. It covers the whole area in which the attraction or repulsion of a magnet can be felt. Magnetic fields such as that of Earth cause magnetic compass needles and other permanent magnets to line up in the direction of the field.

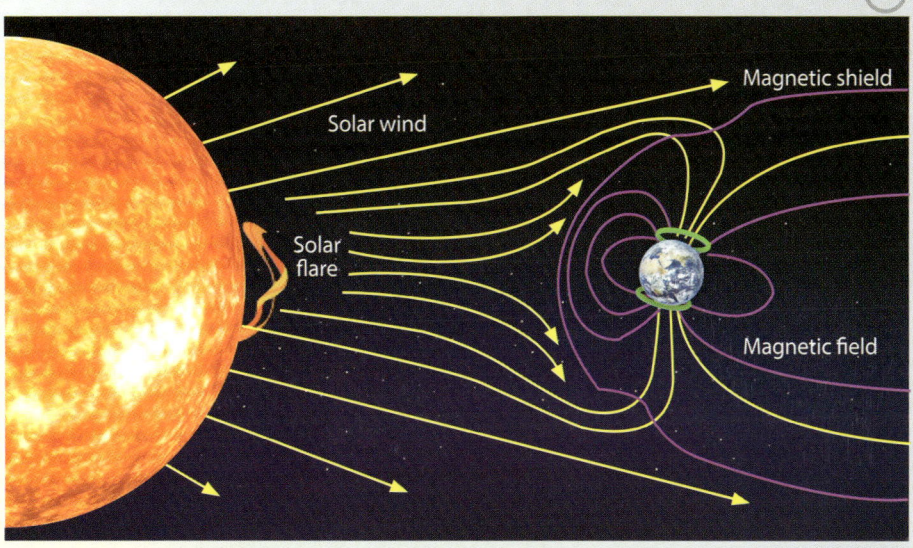
▲ How auroras are caused by the Sun

Sometimes, due to **solar winds** and storms, the amount of energy sent by the Sun varies. During one particular type of solar storm, the Sun ejects a large bubble of electrified gas which can travel at great speeds through space. When a solar storm like this approaches us on Earth, the energy and small particles move down into Earth's atmosphere along the magnetic field lines at the North Pole and South Pole.

When the particles reach Earth's atmosphere, they engage with the gases in it, and this results in amazing displays of bright lights in the sky. Dazzling green and red lights are seen due to the interaction with oxygen, whereas nitrogen gives off blue and purple hues.

◀ The Southern lights in Tasmania

▲ Aurora Australis over Lake Wakatipu, South Island, New Zealand

Extrasolar Planets or Exoplanets

For a long time, human beings only knew about the existence of planets within the solar system. It was only in 1992 that planets outside our solar system were also discovered. They were called extrasolar planets or exoplanets. Astronomers now know that exoplanets are a usual occurrence in the universe, and they are aware of more than 3,000 exoplanets. They are also in the process of getting information about an additional 1,000 or more.

What are Exoplanets?

An extrasolar or exoplanet is a planetary body that is outside the solar system and usually orbits a star other than the Sun, unlike the planets in our solar system, which orbit the Sun. They are very hard to view even with telescopes, since they are generally concealed due to the bright glare of the stars which they orbit.

◀ This illustration shows a comparison between the solar system and the Kepler-22 System. Keppler-22b is an exoplanet and Keppler-22 is the star around which it revolves

▲ An illustration of a faraway planet system with exoplanets (elements of this image are furnished by NASA)

▼ Illustration of an exoplanet, with exo-moons orbiting a binary star system (elements of this image are furnished by NASA)

In Real Life

NASA launched the Kepler spacecraft in 2009 to find exoplanets. Kepler searched for such planets varying in size and orbit, and also those whose stars differed in size and temperature. Some exoplanets found by Kepler appear to be rock-like and at unusual distances from their stars. Such exoplanets seem to have a habitable area and may support life. The Kepler space mission has discovered thousands of extrasolar planets.

In 2017, NASA discovered seven Earth-sized planets in the habitable zone of TRAPPIST-1, a single star. All the seven rocky planets seem to have the possibility of having water on their surface. It is a crucial discovery for scientists in their search for life on other planets. After our solar system and planets, scientists know the most about the TRAPPIST-1 planetary system.

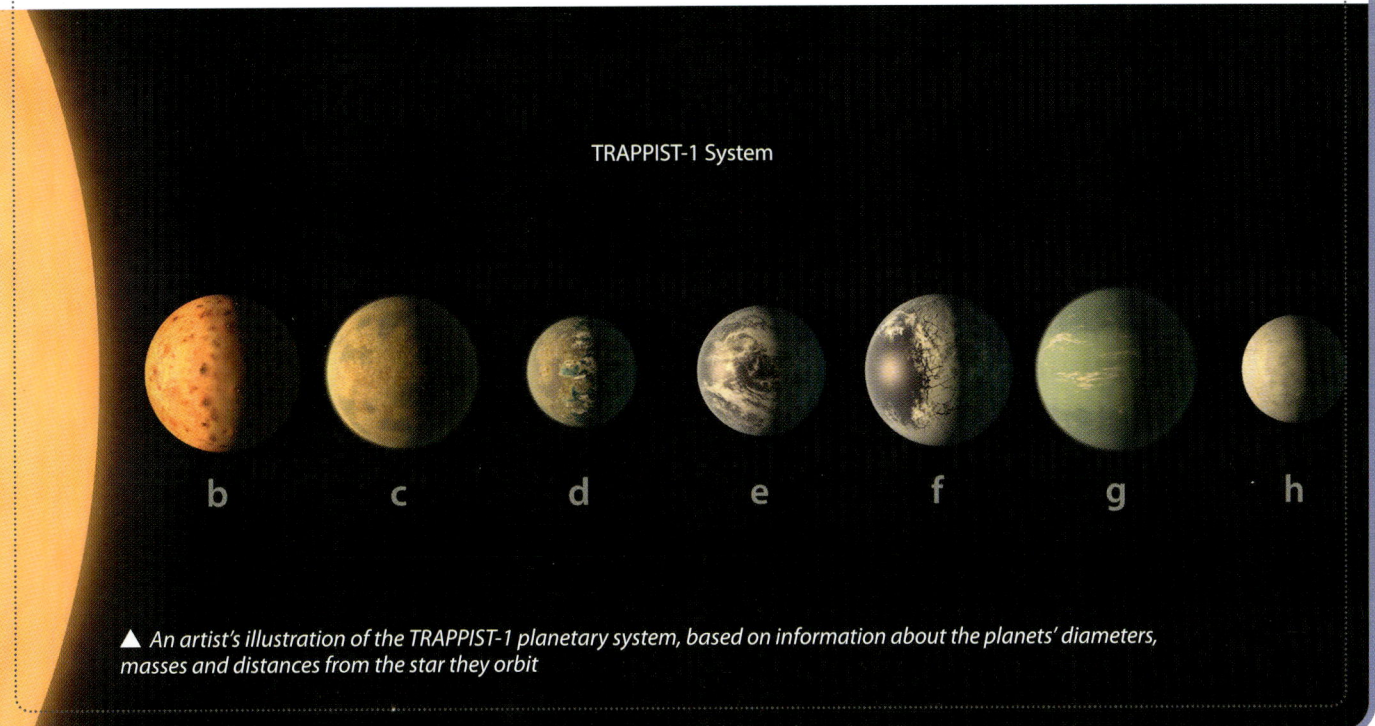

▲ An artist's illustration of the TRAPPIST-1 planetary system, based on information about the planets' diameters, masses and distances from the star they orbit

How can we Spot Exoplanets?

One method to discover exoplanets is to look for unsteady or 'wobbly' stars. Stars which have planets orbiting them tend to appear shaky or wobbly since they do not orbit perfectly around the star's centre. Due to this off-centre orbit, when viewed from a long distance, the star which has planets around it looks like it is wobbling.

While numerous exoplanets have been discovered using this method, it is suitable for the discovery of only large planets like Jupiter, or those larger than Jupiter. Small planets the size of Earth or smaller are more difficult to find using this method, since the wobble is so small that it is difficult to detect.

Another method used by scientists to discover exoplanets is the transit method. The revolution of a planet around a star is known as a transit. When this happens, the planet blocks out some of the star's light. So, a star will look a bit dimmer when the planet passes in front of it. By observing the changes in the brightness of the star during transit, astronomers can figure out the size of the planet. Observing the time taken between transits tells scientists how far the planet is from its star. If the temperature is ideal or right, the planet may have a **habitable zone**.

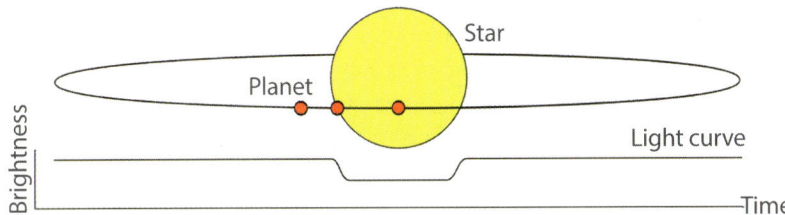

▲ The image shows a planet transiting a star—the 'transit method' is used to find extrasolar planets

Universe or Multiverse?

We have always thought of our universe as unique and one of a kind. In fact, the word itself implies this. Based on an enhanced version of the Big Bang model, a group of four physicists in the 1980s first came up with the idea that there may be other universes or multiverses that also exist.

▼ *The universe comprises stars, galaxies and nebulas, among other celestial bodies*

⭐ Multiverse Theory

The term 'multiverse' was first mentioned by American philosopher William James in 1895. There are four physicists, however, who are responsible for suggesting the theory of inflation. They said that when the Big Bang took place nearly 14 billion years ago, within the first few seconds, the universe underwent an expansion, or what is termed as 'inflation'. Satellite measurements of the heat which was left behind by the Big Bang also support this theory. It suggested something quite out of the ordinary. According to this theory, the Big Bang that created the universe may not have been a one-time event, but rather it may have occurred again and again an unending number of times. These scientists claimed that each of these big bangs would have possibly created other universes. Therefore, they concluded that rather than one universe, we may be living in a multiverse (many universes).

▲ *American philosopher and psychologist William James (1842–1910) was a professor at Harvard College for several years*

Mathematic calculations revealed that the multiverse may comprise of a variety of universes, some very different from our universe and others with replicas of Earth, including all of us in it. The existence of other universes would also mean that things that can happen in one universe can also happen in the other universes. It is a debatable and controversial idea.

Hence, according to the multiverse theory, our universe, which stretches about 90 billion light years across, would then form a very small part of the multiverse. The multiverse idea has been discussed in the study of cosmology, quantum mechanics and also philosophy.

Units of Measurement, and Cosmic Forces

How do astronomers and scientists measure distances in the universe? What, in fact, holds our universe together? Read on to find out.

Astronomical Unit

An astronomical unit (or AU) is a unit of distance. It is approximately the average distance between Earth and the Sun. So, one AU is about 15,00,00,000 kilometres.

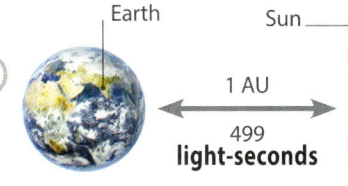

▲ *An illustration depicting an astronomical unit*

Light Year

A light year is the distance travelled by light in one Earth year. One light year is approximately 9 trillion kilometres (9 with 12 zeroes behind it). It is used to describe the distance of objects in space.

▲ *The illustration shows the pull of gravity, which makes objects fall downwards instead of floating in the air*

Nuclear Fusion

Nuclear fusion is a process by which nuclear reactions between light elements form heavier elements (up to iron). The nuclear fusion process within the core of a star produces a lot of energy, which travels into space in the form of heat and light.

Gravity

When you throw an apple into the air, why does it fall down instead of floating away into the sky? This is due to **gravity,** a universal and invisible force by which a planet or body attracts objects towards its centre. We would not be able to exist on Earth without gravity; it keeps everything in the universe in place, including planets and even Earth's atmosphere.

What Exerts Gravity?

All objects that have mass have gravity, including human beings. The more the mass, the greater is the gravity. Earth's gravity is possible due to all of its mass, which creates a gravitational pull on the entire mass in your body. This gives you weight. But if you were on a planet or body which had lesser mass than Earth, you would weigh much less and vice versa!

Isn't It Amazing!

When you stand on a weighing scale, it actually measures how hard Earth's gravity is pulling you. On Mercury or Jupiter, it would show you a different figure. Since the planets differ in their weights, the gravity they exert on us is also different.

If you weigh 45 kilograms on Earth, you would weigh only 17 kilograms on Mercury and about 115 kilograms on Jupiter. This is because the weight of Mercury is less than Earth and, therefore, its gravity would pull lesser on your body, whereas Jupiter weighs more, so the pull on your body would be much greater!

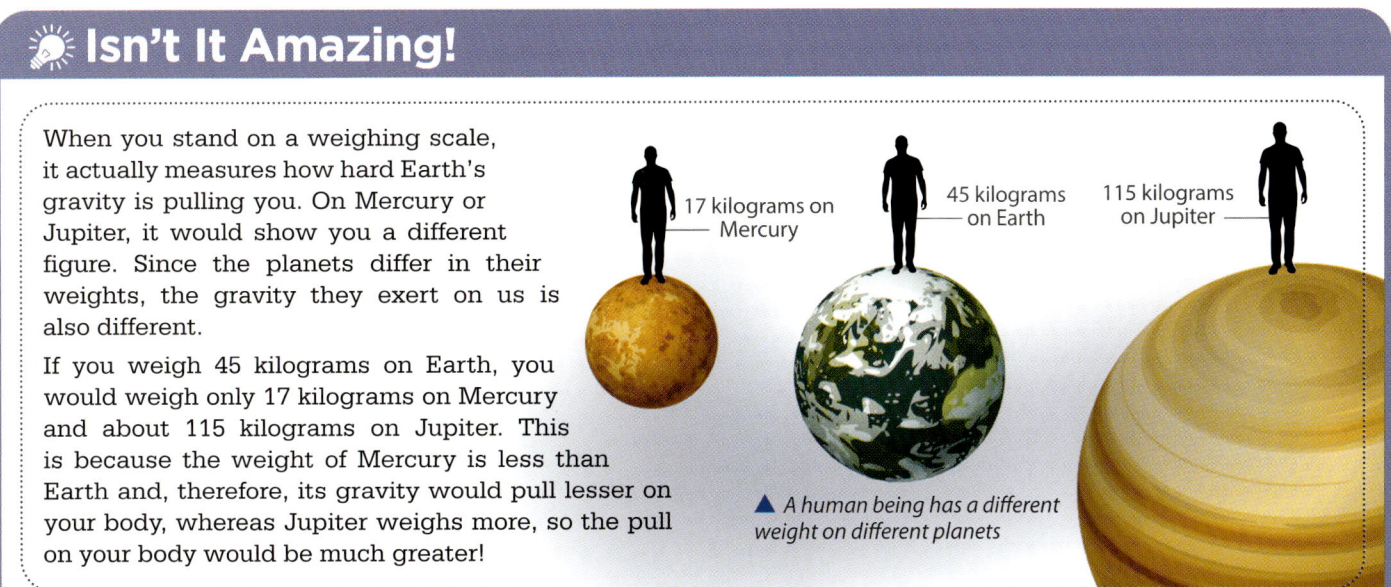

▲ *A human being has a different weight on different planets*

Satellites
Natural & Artificial

We are all familiar with the word 'satellite' and we know that there are both natural and artificial satellites. Natural satellites, like Earth's Moon, are not created by people, unlike artificial satellites. Natural satellites like our Moon can exist in a variety of shapes, sizes and types. They are generally solid bodies and very few of them have atmospheres. They were most probably created from the gas and dust moving around planets in the early days of the solar system. Artificial satellites, on the other hand, are made by human beings and have some specific uses.

What is a Satellite?

A moon, planet or a machine that orbits a planet or star is known as a satellite. Our Earth orbits the Sun and the Moon orbits Earth, hence both are examples of natural satellites.

The word 'satellite' is more commonly used to refer to machines that are launched into space and move around Earth or any other celestial body in space. These are man-made or artificial satellites. They aid scientists in their studies and help them get more information about our universe, solar system, oceans, land, and atmosphere. Artificial satellites are also useful since they are able to take photographs of deep space objects and other phenomena and send them back to Earth. For example, pictures of our planet taken from space help meteorologists forecast the weather and predict natural disasters like hurricanes.

Other types of satellites are used for communication purposes, like those which beam TV and phone signals across the world or those which help us get important and useful information like our exact location.

◀ *The Moon is Earth's natural satellite*

Moons of the Solar System

Our solar system consists of hundreds of moons. Some asteroids have also been found to have small moons. While Earth has only one moon or satellite, some other planets have many more, while some planets have none.

The two planets closest to the Sun are Mercury and Venus. They do not have any moons. Mercury is so very close to the Sun and its gravity, that it would, most probably, not be able to hold on to its moon. It would either crash into it or get sucked up by the Sun.

Mars, on the other hand, has two moons. Jupiter, the outer giant planet can boast of 79 moons (53 of them have names, while the others are yet to be officially named)! It also has the biggest moon in the solar system, called Ganymede. Jupiter's moons are so large that they can be viewed with a pair of binoculars on a clear, dark night.

Saturn currently has 53 named moons and may have another nine which are still to be confirmed. If they are confirmed, Saturn will have 62 moons in all. Of the 27 moons which are known so far to be orbiting around Uranus, some are partially made of ice. Neptune, till date, has been found to have 13 moons and may have one more, but it is not yet confirmed.

▶ *The illustration shows the moons of the solar system and their comparative sizes to each other and to Earth*

▲ Uranus with some of its moons

 ## Moons of the Kuiper Belt

The Kuiper Belt, which surrounds the solar system and is the region where Pluto lies, has several planets and moons. Pluto has a fairly large moon, Charon, which is half the size of Pluto. NASA's Hubble Space Telescope found two more very tiny moons orbiting Pluto in 2005. Two other dwarf planets Eris and Haumea have one and two moons each.

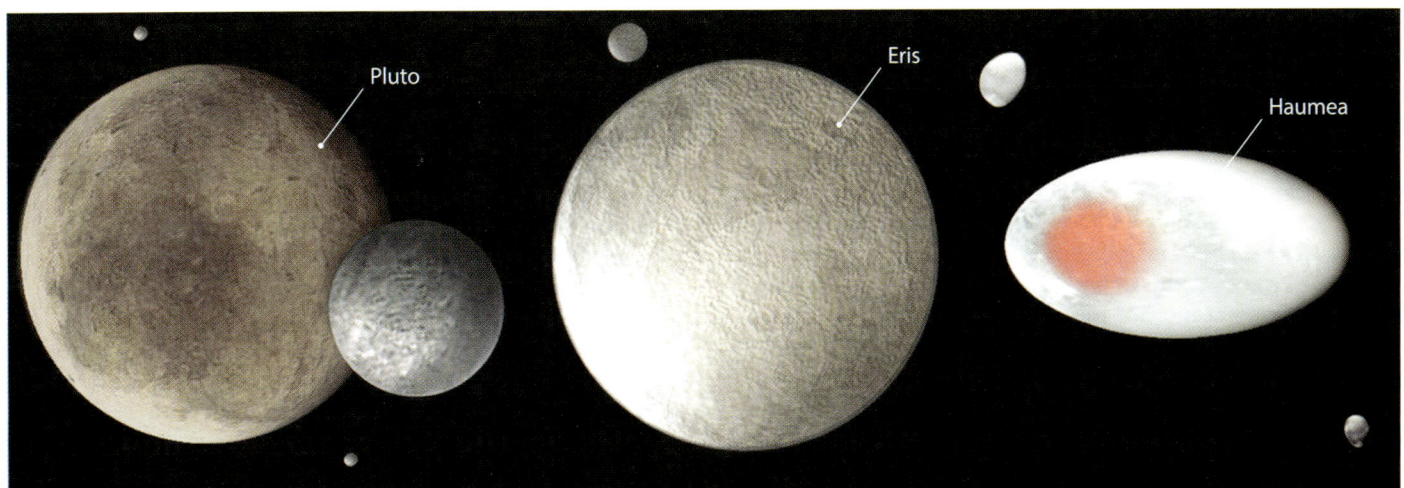

▲ Dwarf planets and their moons

Concepts about the Universe

Cosmology is the scientific study of the evolution of the universe. Cosmologists are interested in understanding the past, present and the future of the universe. For centuries, human beings have been trying to understand the cosmos. Early concepts about the universe were vastly different from modern-day theories and have undergone several changes.

⭐ Early Concepts of the Universe

Greek philosopher and scientist Aristotle (384–322 BCE) was one of the greatest scholars of the Western world. He proposed that Earth was the centre of the universe, with the Sun, Moon and planets, as well as the fixed stars revolving around it. Aristotle's views about the universe were accepted by a majority of the Greeks in his time.

300 years later, Ptolemy of Alexandria was the first astronomer to make scientific maps of the skies. But he too believed in the Earth-centred view of the universe. This was the major belief for about 1500 years.

▲ The great Greek philosopher and scientist Aristotle

▶ Greek-born Egyptian astronomer, mathematician and geographer Ptolemy (100–170 CE)

⭐ The Copernican Revolution

Nicolaus Copernicus, a Polish monk, challenged this Earth-centred view of the universe in 1543. He suggested that the Sun was at the centre of the universe. Since the Catholic Church believed that Earth was the central figure, Copernicus's theory was met with a lot of hostility. The views and precise studies of Tycho Brahe, as well as Italian scientist Galileo Galilei's use of the telescope, eventually led to Copernicus's theory being accepted. In 1609, Galilei developed his own telescope and used it to study the skies. He observed a smaller version of the solar system, as suggested by Copernicus, with moons travelling around the planets. His discoveries revolutionised the science of astronomy.

◀ Nicolaus Copernicus proposed the heliocentric or Sun-centred solar system, in which Earth and the other planets move around the Sun

⭐ Incredible Individuals

In 1633, Galilei was convicted by the Roman Catholic Church for supporting the Copernican theory, according to which Earth revolves around the Sun. The Church criticised him for going against the sacred writings of the Bible as they considered Earth and not the Sun as the centre of the universe. They asked him to withdraw his conclusions and put him on a 13-year trial and investigation. He was also put under house arrest for nearly eight years before he died in 1642.

Ironically, more than 350 years after the Roman Catholic Church condemned Galilei, in 1992, the Vatican issued a note to say that he was right after all and that Earth is not fixed, but it does, in fact, move around the Sun!

▶ Galileo Galilei

Modern Astronomy and Science

Both Galilei and British scientist Sir Isaac Newton laid the foundations of modern science. Newton used mathematics to explain known facts and also made mathematical laws to explain how objects moved in space and on Earth. He explained orbiting planets and concluded that all celestial bodies are always moving, with no limits on space and time. In 1917, scientist Albert Einstein described the universe based on his theory of general relativity. According to it, time passes more slowly for objects in gravitational fields (like for us on Earth) than for objects far from such fields.

▲ Two great men: (Top) Galileo Galilei (1564–1642) explaining how to use the telescope to the Doge of Venice; and (right) Sir Isaac Newton (1643–1727) in his lab, explaining an optical experiment

Galilei made major contributions to the sciences of motion, astronomy and the development of scientific method. Newton was a key person during the scientific revolution of the 17th century. He is best known for his three laws of motion, amongst numerous other discoveries and scientific formulations.

Einstein's theory influenced many scientists. Two of them were Willem de Sitter of Holland and Aleksandr Friedmann from Russia. A lot of cosmology today is based on Friedmann's expansion of Einstein's equations of general relativity, which helped gain a better understanding of the evolution of the universe.

▶ German-born Albert Einstein (1879–1955) was a world-renowned physicist known for developing the special and general theories of relativity. He won the Nobel Prize for Physics in 1921 and is recognised as the most influential physicist of the 20th century

Another important discovery was made by American astronomer Edwin Hubble in the 1920s. For hundreds of years, astronomers thought that the universe only consisted of the Milky Way galaxy. Hubble was the first to observe the existence of other galaxies that were not part of the Milky Way. He concluded that galaxies were moving away from us at great speeds and that the universe was expanding.

◀ American astronomer Edwin Hubble (1889–1953) was a key figure in establishing the field of extragalactic astronomy. He is considered as the leading observational cosmologist of the 20th century

Word Check

Antiparticle: It is a subatomic particle that has similar mass to a particle, but it has opposite electric or magnetic properties.

Atoms: They are the basic units of matter and the defining structure of elements.

Big Bang model: It is the theory of the evolution of the universe that states that the universe emerged from a state of extremely high temperature and density, which occurred around 13.8 billion years ago.

Chemical elements: They are substances that cannot be decomposed into simpler substances by ordinary chemical processes. Elements are the fundamental materials of which all matter is composed.

Cosmic microwave background: It is electromagnetic radiation filling the universe and is a residual effect of the Big Bang.

Cosmology: It is the science of the beginning and evolution of the universe. It includes the study of theories like the Big Bang model along with astronomical subjects and particle physics.

Electron: A subatomic particle with a negative charge commonly found in the outer layers of an atom.

Fjord: It is a long and narrow area of sea lying between high cliffs found in countries like Norway.

Gravity: Also referred to as gravitation, it is a universal force by which a planet or other body attracts objects towards its centre.

Habitable zone: It is the orbital region around a star in which an Earth-like planet can possess water on its surface and possibly support life.

Interstellar: It is the region between the stars that contains vast, diffuse clouds of gases and minute solid particles.

Light-second: It refers to the distance travelled by light in one second, in a vacuum. It amounts to approximately 3,00,000 kilometres.

Light year: It is the distance travelled by light in one year, at a speed of 3,00,000 kilometres per second.

Molecule: It is a group of two or more atoms that forms the smallest identifiable unit into which a pure substance can be divided and still retain the composition and chemical properties of that substance.

Neutron: A subatomic particle with zero charge, commonly found in the nucleus of an atom

Protons: A subatomic particle with a positive charge, commonly found in the nucleus of an atom

Radiation: It is the emission of energy in the form of electromagnetic waves or moving subatomic particles, especially high-energy particles causing ionisation.

Radio waves: They are electromagnetic waves of certain frequency used for long-distance communication.

Solar wind: A continuous outflow of solar subatomic particles from the outer regions of the Sun into the solar system

Singularity: In reference to a black hole, when a star dies, the crushing weight of constituent matter falling in from all sides compresses the dying star to a point of zero volume and infinite density called the singularity

Volume: It is the amount of space that is occupied by any substance or object.